The Elves and the Shoemaker

Gaynor Bee, Dav
and Wendy P

GW00371687

Illustrated by Chris Rothero

Cambridge University Press

Cambridge   New York   Port Chester   Melbourne   Sydney

Once upon a time there lived an old shoemaker and his wife. The old man made shoes better than any other shoemaker in town. But he worked very slowly and carefully. The old couple didn't sell many shoes in a week. So they were very poor.

One day the shoemaker had just one piece of leather
left and no money to buy more. It took him a long time
to cut out the leather for the shoes. "This may be the
last pair of shoes I ever make," he said to his wife. That
night he left the cut out pieces on his bench and went
sadly to bed.

But when he came down to his workshop next morning, there was a surprise waiting for him. On his bench was the finest pair of shoes he had ever seen. "What beautiful stitches! These shoes will sell for twice the usual price," he thought. And they did.

Now there was enough money to buy leather to make two pairs of shoes. The shoemaker cut out the shoes and then left the pieces on the bench and went to bed.

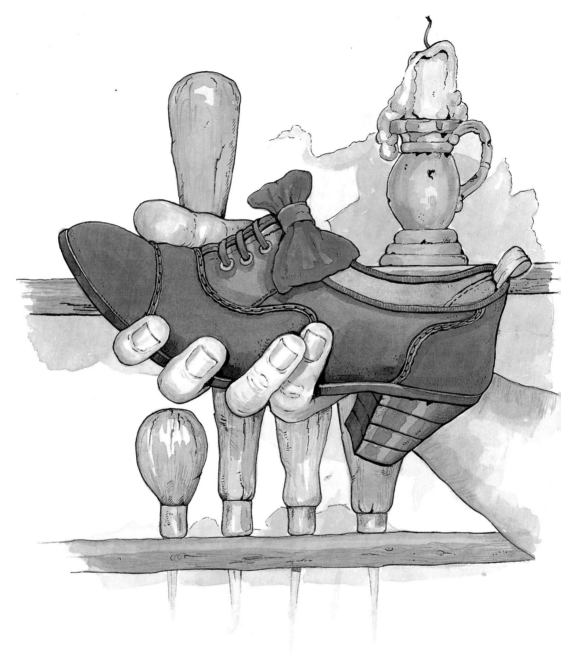

When the shoemaker came down the next morning, there on the bench were two new pairs of shoes. Each pair was as beautifully made as the pair he'd found the day before.

Almost as soon as his wife put the new shoes in the
window they were sold. The old woman used some of
the money to buy more leather.

The next morning he found yet more beautiful shoes on his bench. And so it was every morning after that. The old couple wondered who was making the shoes, but they were afraid to spy on them in case they frightened their magical helpers away.

The shoemaker's shop became the most famous in town. People came from far and wide to buy the shoes and the old man and his wife soon had enough money to live comfortably.

"We must find out who makes the shoes," said the old
woman. So one night they hid and watched what
happened in the workshop.

At midnight six tiny elves climbed onto the workbench and set about making shoes. To their surprise, the old couple noticed that the elves' clothes were so frayed and torn that they shivered with cold. "Poor little fellows," said the old woman. "They work so hard, we must give them some thank you presents."

So in the morning the old couple set to work making
tiny boots and clothes for the elves, to keep them
warm at night.

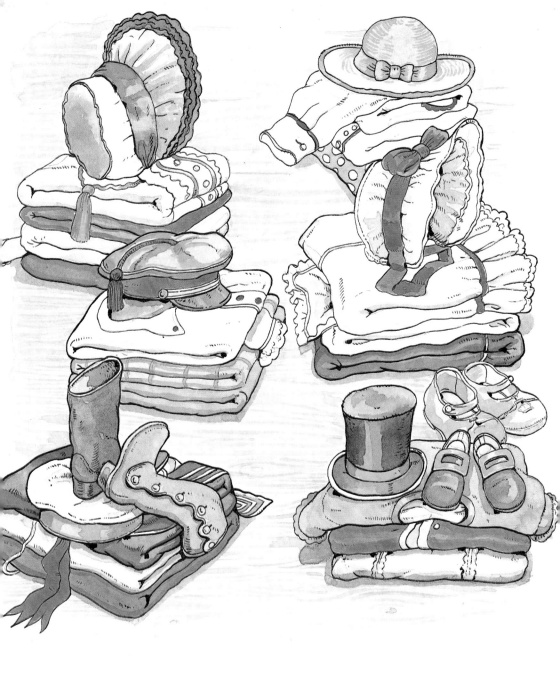

That night they left the tiny clothes and boots on the
bench and hid again.

This time, when the elves arrived they didn't find any
leather to sew. All they saw was their new clothes.
They pulled them on and danced around the
workbench. They were laughing as they skipped away
down the street.

The old shoemaker and his wife smiled as they went up
to bed.
"They know we don't need their help any more," said
the old man. "Tomorrow we can open up the shop and
sell more shoes."

Activities

Using construction kits design and make the shoemaker's shop. Show where the shoes are put on show. The old man and woman also live in the shop. Show where they live. The shop was part of a street in a town. Design and make the street in the town.

Once the shoemaker and his wife had earned some extra money they needed a new sign outside the shop. The people who passed by needed to know what was sold in the shop. Design and make a sign that would attract people's attention and make them look in the window to see the beautiful shoes.

Rearrange the home corner to become the shoemaker's shop and home. Where are the shoes going to be made? Where will the beautiful shoes be displayed for sale? Where will the shoemaker and his wife live? Where will the elves live?

Published by the Press Syndicate of the University of Cambridge
The Pitt Building, Trumpington Street, Cambridge CB2 1RP
10 Stamford Road, Oakleigh, Melbourne 3166, Australia

First published 1991

In association with
Staffordshire County Council

Designed and phototypeset by Gecko Limited, Bicester, Oxon

Printed in Great Britain at the University Press, Cambridge

This book is only available as part of a pack

ISBN 0 521 40617 X